THIS BOOK BELONGS TO:

CONTACT INFORMATION	
NAME:	
ADDRESS:	
PHONE:	

Copyright © Teresa Rother

All rights reserved. No part of this publication may be reproduced, distributed, or transmitted in any form or by any means, including photocopy, recording, or other electronic or mechanical methods.

DEDICATION

This Goat Record Keeping Book is dedicated to goat owners, small breeders, farmers, and ranchers who want to monitor and record important information regarding their goat herd.

You are my inspiration for producing this book and I'm honored to be a part of helping you retain and collect data of your goat herd.

HOW TO USE THIS BOOK

This Goat Recording Keeping Book will help record, collect, and organize your information in an easy to use format.

Here are examples of information for you to fill in and write the details of your recipe collection.

Fill in the following information:

1. Goat Information - name, breed, birth date, identification details, medical information, pedigree chart, a place for notes

2. Medical Information - injury or illness, parasite control, testing record, vaccination, and supplement record

3. Doe's Kidding Record - name, date, breed to, number of kids, sex (buck or doe), name of kid, sire, weight, tattoo

4. Buck's Record Of Progeny - name, year, breed to, buck or doe

5. Goat Record - name, breed, date of birth, date weened, identification, weight by month, feed record by month (grain, field or pasture)

6. Milk Production - monthly and yearly record

GOAT INFORMATION

NAME		BUCK	DOE
BREED		BIRTH DATE:	
DATE ACQUIRED:	HOW ACQUIRED: BORN ON FARM ☐ PURCHASED ☐ LEASED ☐		
COLORS / IDENTIFYING MARKS:			
PURPOSE:	MILK ☐ MEAT ☐ PET ☐	OTHER ☐	
NOTES			

PEDIGREE CHART

- SIRE
  - GRAND SIRE
    - GREAT GRAND SIRE
    - GREAT GRAND DAM
  - GRAND DAM
    - GREAT GRAND SIRE
    - GREAT GRAND DAM
- DAM
  - GRAND SIRE
    - GREAT GRAND SIRE
    - GREAT GRAND DAM
  - GRAND DAM
    - GREAT GRAND SIRE
    - GREAT GRAND DAM

MEDICAL INFORMATION

INJURY OR ILLNESS

DATE	DESCRIPTION OR NATURE OF ILLNESS	TREATMENT

PARASITE CONTROL

DATE	METHOD OR DEWORMER	DATE	METHOD OR DEWORMER

TESTING RECORD

DATE	TEST PERFORMED (CAE, CL, TB...)	RESULT	DATE	TEST PERFORMED (CAE, CL, TB...)	RESULT

VACCINATION & SUPPLEMENT RECORD

DATE	TARGET DISEASE	DRUG OR SUPPLEMENT USED	DOSAGE	RESULTS

DOE'S KIDDING RECORD

DOE'S NAME:

DATE BREED	KIDDING DATE	# OF KIDS	SEX D/B	NAME OF KID	SIRE OF KID	WEIGHT	TATTOO

BUCK'S RECORD OF PROGENY

BUCK'S NAME:			

YEAR	BRED TO	KIDS	DOE/BUCK

GOAT RECORD

GOAT'S NAME:		IDENTIFICATION:
BREED:	DATE OF BIRTH:	DATE OF WEANED:

WEIGHT (POUNDS)

BIRTH	JAN	FEB	MAR	APR	MAY	JUN	JUL	AUG	SEP	OCT	NOV	DEC	FINAL

FEED RECORD

	JAN	FEB	MAR	APR	MAY	JUN	JUL	AUG	SEP	OCT	NOV	DEC	TOTAL
GRAIN													
FIELD													
PASTURE													

MILK PRODUCTION

DOE'S NAME:		IDENTIFICATION:	
BREED:	DATE OF BIRTH:	KIDDING DATE:	

JANUARY		AVERAGE LBS / DAY X 31 DAYS =		LBS
FEBRUARY		AVERAGE LBS / DAY X 31 DAYS =		LBS
MARCH		AVERAGE LBS / DAY X 31 DAYS =		LBS
APRIL		AVERAGE LBS / DAY X 31 DAYS =		LBS
MAY		AVERAGE LBS / DAY X 31 DAYS =		LBS
JUNE		AVERAGE LBS / DAY X 31 DAYS =		LBS
JULY		AVERAGE LBS / DAY X 31 DAYS =		LBS
AUGUST		AVERAGE LBS / DAY X 31 DAYS =		LBS
SEPTEMBER		AVERAGE LBS / DAY X 31 DAYS =		LBS
OCTOBER		AVERAGE LBS / DAY X 31 DAYS =		LBS
NOVEMBER		AVERAGE LBS / DAY X 31 DAYS =		LBS
DECEMBER		AVERAGE LBS / DAY X 31 DAYS =		LBS
YEARLY TOTAL MILK PRODUCED =				LBS
TOTAL VALUE OF MILK PRODUCED FOR THE YEAR				
	LBS X $		VALUE PER LBS =	

GOAT INFORMATION

PHOTO

NAME		BUCK	DOE
BREED		BIRTH DATE:	
DATE ACQUIRED:	HOW ACQUIRED: BORN ON FARM ☐ PURCHASED ☐ LEASED ☐		
COLORS / IDENTIFYING MARKS:			
PURPOSE:	MILK ☐ MEAT ☐ PET ☐	OTHER ☐	
NOTES			

PEDIGREE CHART

- SIRE
  - GRAND SIRE
    - GREAT GRAND SIRE
    - GREAT GRAND DAM
  - GRAND DAM
    - GREAT GRAND SIRE
    - GREAT GRAND DAM
- DAM
  - GRAND SIRE
    - GREAT GRAND SIRE
    - GREAT GRAND DAM
  - GRAND DAM
    - GREAT GRAND SIRE
    - GREAT GRAND DAM

MEDICAL INFORMATION

INJURY OR ILLNESS

DATE	DESCRIPTION OR NATURE OF ILLNESS	TREATMENT

PARASITE CONTROL

DATE	METHOD OR DEWORMER	DATE	METHOD OR DEWORMER

TESTING RECORD

DATE	TEST PERFORMED (CAE, CL, TB...)	RESULT	DATE	TEST PERFORMED (CAE, CL, TB...)	RESULT

VACCINATION & SUPPLEMENT RECORD

DATE	TARGET DISEASE	DRUG OR SUPPLEMENT USED	DOSAGE	RESULTS

DOE'S KIDDING RECORD

DOE'S NAME:

DATE BREED	KIDDING DATE	# OF KIDS	SEX D/B	NAME OF KID	SIRE OF KID	WEIGHT	TATTOO

BUCK'S RECORD OF PROGENY

BUCK'S NAME:	

YEAR	BRED TO	KIDS	DOE/BUCK

GOAT RECORD

GOAT'S NAME:	IDENTIFICATION:
BREED:	DATE OF BIRTH:	DATE OF WEANED:

WEIGHT (POUNDS)

BIRTH	JAN	FEB	MAR	APR	MAY	JUN	JUL	AUG	SEP	OCT	NOV	DEC	FINAL

FEED RECORD

	JAN	FEB	MAR	APR	MAY	JUN	JUL	AUG	SEP	OCT	NOV	DEC	TOTAL
GRAIN													
FIELD													
PASTURE													

MILK PRODUCTION

DOE'S NAME:		IDENTIFICATION:	
BREED:	DATE OF BIRTH:	KIDDING DATE:	

JANUARY		AVERAGE LBS / DAY X 31 DAYS =		LBS
FEBRUARY		AVERAGE LBS / DAY X 31 DAYS =		LBS
MARCH		AVERAGE LBS / DAY X 31 DAYS =		LBS
APRIL		AVERAGE LBS / DAY X 31 DAYS =		LBS
MAY		AVERAGE LBS / DAY X 31 DAYS =		LBS
JUNE		AVERAGE LBS / DAY X 31 DAYS =		LBS
JULY		AVERAGE LBS / DAY X 31 DAYS =		LBS
AUGUST		AVERAGE LBS / DAY X 31 DAYS =		LBS
SEPTEMBER		AVERAGE LBS / DAY X 31 DAYS =		LBS
OCTOBER		AVERAGE LBS / DAY X 31 DAYS =		LBS
NOVEMBER		AVERAGE LBS / DAY X 31 DAYS =		LBS
DECEMBER		AVERAGE LBS / DAY X 31 DAYS =		LBS
YEARLY TOTAL MILK PRODUCED =				LBS

TOTAL VALUE OF MILK PRODUCED FOR THE YEAR

| | LBS X $ | | VALUE PER LBS = | |

GOAT INFORMATION

NAME		BUCK	DOE
BREED		BIRTH DATE:	
DATE ACQUIRED:	HOW ACQUIRED: BORN ON FARM ☐ PURCHASED ☐ LEASED ☐		
COLORS / IDENTIFYING MARKS:			
PURPOSE: MILK ☐ MEAT ☐ PET ☐ OTHER ☐			
NOTES			

PEDIGREE CHART

- SIRE
  - GRAND SIRE
    - GREAT GRAND SIRE
    - GREAT GRAND DAM
  - GRAND DAM
    - GREAT GRAND SIRE
    - GREAT GRAND DAM
- DAM
  - GRAND SIRE
    - GREAT GRAND SIRE
    - GREAT GRAND DAM
  - GRAND DAM
    - GREAT GRAND SIRE
    - GREAT GRAND DAM

MEDICAL INFORMATION

INJURY OR ILLNESS

DATE	DESCRIPTION OR NATURE OF ILLNESS	TREATMENT

PARASITE CONTROL

DATE	METHOD OR DEWORMER	DATE	METHOD OR DEWORMER

TESTING RECORD

DATE	TEST PERFORMED (CAE, CL, TB...)	RESULT	DATE	TEST PERFORMED (CAE, CL, TB...)	RESULT

VACCINATION & SUPPLEMENT RECORD

DATE	TARGET DISEASE	DRUG OR SUPPLEMENT USED	DOSAGE	RESULTS

DOE'S KIDDING RECORD

DOE'S NAME:

DATE BREED	KIDDING DATE	# OF KIDS	SEX D/B	NAME OF KID	SIRE OF KID	WEIGHT	TATTOO

BUCK'S RECORD OF PROGENY

BUCK'S NAME:	

YEAR	BRED TO	KIDS	DOE/BUCK

GOAT RECORD

GOAT'S NAME:	IDENTIFICATION:
BREED:	DATE OF BIRTH:	DATE OF WEANED:

WEIGHT (POUNDS)

BIRTH	JAN	FEB	MAR	APR	MAY	JUN	JUL	AUG	SEP	OCT	NOV	DEC	FINAL

FEED RECORD

	JAN	FEB	MAR	APR	MAY	JUN	JUL	AUG	SEP	OCT	NOV	DEC	TOTAL
GRAIN													
FIELD													
PASTURE													

MILK PRODUCTION

DOE'S NAME:		IDENTIFICATION:	
BREED:	DATE OF BIRTH:	KIDDING DATE:	

Month		Calculation		LBS
JANUARY		AVERAGE LBS / DAY X 31 DAYS =		LBS
FEBRUARY		AVERAGE LBS / DAY X 31 DAYS =		LBS
MARCH		AVERAGE LBS / DAY X 31 DAYS =		LBS
APRIL		AVERAGE LBS / DAY X 31 DAYS =		LBS
MAY		AVERAGE LBS / DAY X 31 DAYS =		LBS
JUNE		AVERAGE LBS / DAY X 31 DAYS =		LBS
JULY		AVERAGE LBS / DAY X 31 DAYS =		LBS
AUGUST		AVERAGE LBS / DAY X 31 DAYS =		LBS
SEPTEMBER		AVERAGE LBS / DAY X 31 DAYS =		LBS
OCTOBER		AVERAGE LBS / DAY X 31 DAYS =		LBS
NOVEMBER		AVERAGE LBS / DAY X 31 DAYS =		LBS
DECEMBER		AVERAGE LBS / DAY X 31 DAYS =		LBS
YEARLY TOTAL MILK PRODUCED =				LBS

TOTAL VALUE OF MILK PRODUCED FOR THE YEAR			
	LBS X $		VALUE PER LBS =

GOAT INFORMATION

PHOTO

NAME		BUCK	DOE
BREED		BIRTH DATE:	
DATE ACQUIRED:	HOW ACQUIRED: BORN ON FARM ☐ PURCHASED ☐ LEASED ☐		
COLORS / IDENTIFYING MARKS:			
PURPOSE: MILK ☐ MEAT ☐ PET ☐ OTHER ☐			
NOTES			

PEDIGREE CHART

- SIRE
  - GRAND SIRE
    - GREAT GRAND SIRE
    - GREAT GRAND DAM
  - GRAND DAM
    - GREAT GRAND SIRE
    - GREAT GRAND DAM
- DAM
  - GRAND SIRE
    - GREAT GRAND SIRE
    - GREAT GRAND DAM
  - GRAND DAM
    - GREAT GRAND SIRE
    - GREAT GRAND DAM

MEDICAL INFORMATION

INJURY OR ILLNESS

DATE	DESCRIPTION OR NATURE OF ILLNESS	TREATMENT

PARASITE CONTROL

DATE	METHOD OR DEWORMER	DATE	METHOD OR DEWORMER

TESTING RECORD

DATE	TEST PERFORMED (CAE, CL, TB...)	RESULT	DATE	TEST PERFORMED (CAE, CL, TB...)	RESULT

VACCINATION & SUPPLEMENT RECORD

DATE	TARGET DISEASE	DRUG OR SUPPLEMENT USED	DOSAGE	RESULTS

DOE'S KIDDING RECORD

DOE'S NAME:

DATE BRED	KIDDING DATE	# OF KIDS	SEX D/B	NAME OF KID	SIRE OF KID	WEIGHT	TATTOO

BUCK'S RECORD OF PROGENY

BUCK'S NAME:	

YEAR	BRED TO	KIDS	DOE/BUCK

GOAT RECORD

GOAT'S NAME:	IDENTIFICATION:
BREED:	DATE OF BIRTH:	DATE OF WEANED:

WEIGHT (POUNDS)

BIRTH	JAN	FEB	MAR	APR	MAY	JUN	JUL	AUG	SEP	OCT	NOV	DEC	FINAL

FEED RECORD

	JAN	FEB	MAR	APR	MAY	JUN	JUL	AUG	SEP	OCT	NOV	DEC	TOTAL
GRAIN													
FIELD													
PASTURE													

MILK PRODUCTION

DOE'S NAME:		IDENTIFICATION:	
BREED:	DATE OF BIRTH:	KIDDING DATE:	

JANUARY		AVERAGE LBS / DAY X 31 DAYS =		LBS
FEBRUARY		AVERAGE LBS / DAY X 31 DAYS =		LBS
MARCH		AVERAGE LBS / DAY X 31 DAYS =		LBS
APRIL		AVERAGE LBS / DAY X 31 DAYS =		LBS
MAY		AVERAGE LBS / DAY X 31 DAYS =		LBS
JUNE		AVERAGE LBS / DAY X 31 DAYS =		LBS
JULY		AVERAGE LBS / DAY X 31 DAYS =		LBS
AUGUST		AVERAGE LBS / DAY X 31 DAYS =		LBS
SEPTEMBER		AVERAGE LBS / DAY X 31 DAYS =		LBS
OCTOBER		AVERAGE LBS / DAY X 31 DAYS =		LBS
NOVEMBER		AVERAGE LBS / DAY X 31 DAYS =		LBS
DECEMBER		AVERAGE LBS / DAY X 31 DAYS =		LBS
YEARLY TOTAL MILK PRODUCED =				LBS

TOTAL VALUE OF MILK PRODUCED FOR THE YEAR

	LBS X $		VALUE PER LBS =	

GOAT INFORMATION

NAME		BUCK	DOE
BREED		BIRTH DATE:	
DATE ACQUIRED:	HOW ACQUIRED: BORN ON FARM ☐ PURCHASED ☐ LEASED ☐		
COLORS / IDENTIFYING MARKS:			
PURPOSE:	MILK ☐ MEAT ☐ PET ☐	OTHER ☐	
NOTES			

PEDIGREE CHART

- SIRE
  - GRAND SIRE
    - GREAT GRAND SIRE
    - GREAT GRAND DAM
  - GRAND DAM
    - GREAT GRAND SIRE
    - GREAT GRAND DAM
- DAM
  - GRAND SIRE
    - GREAT GRAND SIRE
    - GREAT GRAND DAM
  - GRAND DAM
    - GREAT GRAND SIRE
    - GREAT GRAND DAM

MEDICAL INFORMATION

INJURY OR ILLNESS

DATE	DESCRIPTION OR NATURE OF ILLNESS	TREATMENT

PARASITE CONTROL

DATE	METHOD OR DEWORMER	DATE	METHOD OR DEWORMER

TESTING RECORD

DATE	TEST PERFORMED (CAE, CL, TB...)	RESULT	DATE	TEST PERFORMED (CAE, CL, TB...)	RESULT

VACCINATION & SUPPLEMENT RECORD

DATE	TARGET DISEASE	DRUG OR SUPPLEMENT USED	DOSAGE	RESULTS

DOE'S KIDDING RECORD

DOE'S NAME:

DATE BREED	KIDDING DATE	# OF KIDS	SEX D/B	NAME OF KID	SIRE OF KID	WEIGHT	TATTOO

BUCK'S RECORD OF PROGENY

BUCK'S NAME:	

YEAR	BRED TO	KIDS	DOE/BUCK

GOAT RECORD

GOAT'S NAME:	IDENTIFICATION:
BREED:	DATE OF BIRTH:	DATE OF WEANED:

WEIGHT (POUNDS)

BIRTH	JAN	FEB	MAR	APR	MAY	JUN	JUL	AUG	SEP	OCT	NOV	DEC	FINAL

FEED RECORD

	JAN	FEB	MAR	APR	MAY	JUN	JUL	AUG	SEP	OCT	NOV	DEC	TOTAL
GRAIN													
FIELD													
PASTURE													

MILK PRODUCTION

DOE'S NAME:		IDENTIFICATION:		
BREED:	DATE OF BIRTH:	KIDDING DATE:		

JANUARY		AVERAGE LBS / DAY X 31 DAYS =		LBS
FEBRUARY		AVERAGE LBS / DAY X 31 DAYS =		LBS
MARCH		AVERAGE LBS / DAY X 31 DAYS =		LBS
APRIL		AVERAGE LBS / DAY X 31 DAYS =		LBS
MAY		AVERAGE LBS / DAY X 31 DAYS =		LBS
JUNE		AVERAGE LBS / DAY X 31 DAYS =		LBS
JULY		AVERAGE LBS / DAY X 31 DAYS =		LBS
AUGUST		AVERAGE LBS / DAY X 31 DAYS =		LBS
SEPTEMBER		AVERAGE LBS / DAY X 31 DAYS =		LBS
OCTOBER		AVERAGE LBS / DAY X 31 DAYS =		LBS
NOVEMBER		AVERAGE LBS / DAY X 31 DAYS =		LBS
DECEMBER		AVERAGE LBS / DAY X 31 DAYS =		LBS
YEARLY TOTAL MILK PRODUCED =				LBS

TOTAL VALUE OF MILK PRODUCED FOR THE YEAR

| | LBS X $ | | VALUE PER LBS = | |

GOAT INFORMATION

PHOTO

NAME		BUCK	DOE
BREED		BIRTH DATE:	
DATE ACQUIRED:	HOW ACQUIRED: BORN ON FARM ☐ PURCHASED ☐ LEASED ☐		
COLORS / IDENTIFYING MARKS:			
PURPOSE: MILK ☐ MEAT ☐ PET ☐	OTHER ☐		
NOTES			

PEDIGREE CHART

- SIRE
  - GRAND SIRE
    - GREAT GRAND SIRE
    - GREAT GRAND DAM
  - GRAND DAM
    - GREAT GRAND SIRE
    - GREAT GRAND DAM
- DAM
  - GRAND SIRE
    - GREAT GRAND SIRE
    - GREAT GRAND DAM
  - GRAND DAM
    - GREAT GRAND SIRE
    - GREAT GRAND DAM

MEDICAL INFORMATION

INJURY OR ILLNESS

DATE	DESCRIPTION OR NATURE OF ILLNESS	TREATMENT

PARASITE CONTROL

DATE	METHOD OR DEWORMER	DATE	METHOD OR DEWORMER

TESTING RECORD

DATE	TEST PERFORMED (CAE, CL, TB...)	RESULT	DATE	TEST PERFORMED (CAE, CL, TB...)	RESULT

VACCINATION & SUPPLEMENT RECORD

DATE	TARGET DISEASE	DRUG OR SUPPLEMENT USED	DOSAGE	RESULTS

DOE'S KIDDING RECORD

DOE'S NAME:

DATE BREED	KIDDING DATE	# OF KIDS	SEX D/B	NAME OF KID	SIRE OF KID	WEIGHT	TATTOO

BUCK'S RECORD OF PROGENY

BUCK'S NAME:	

YEAR	BRED TO	KIDS	DOE/BUCK

GOAT RECORD

GOAT'S NAME:		IDENTIFICATION:	
BREED:	DATE OF BIRTH:		DATE OF WEANED:

WEIGHT (POUNDS)

BIRTH	JAN	FEB	MAR	APR	MAY	JUN	JUL	AUG	SEP	OCT	NOV	DEC	FINAL

FEED RECORD

	JAN	FEB	MAR	APR	MAY	JUN	JUL	AUG	SEP	OCT	NOV	DEC	TOTAL
GRAIN													
FIELD													
PASTURE													

MILK PRODUCTION

DOE'S NAME:		IDENTIFICATION:	
BREED:	DATE OF BIRTH:	KIDDING DATE:	

JANUARY		AVERAGE LBS / DAY X 31 DAYS =		LBS
FEBRUARY		AVERAGE LBS / DAY X 31 DAYS =		LBS
MARCH		AVERAGE LBS / DAY X 31 DAYS =		LBS
APRIL		AVERAGE LBS / DAY X 31 DAYS =		LBS
MAY		AVERAGE LBS / DAY X 31 DAYS =		LBS
JUNE		AVERAGE LBS / DAY X 31 DAYS =		LBS
JULY		AVERAGE LBS / DAY X 31 DAYS =		LBS
AUGUST		AVERAGE LBS / DAY X 31 DAYS =		LBS
SEPTEMBER		AVERAGE LBS / DAY X 31 DAYS =		LBS
OCTOBER		AVERAGE LBS / DAY X 31 DAYS =		LBS
NOVEMBER		AVERAGE LBS / DAY X 31 DAYS =		LBS
DECEMBER		AVERAGE LBS / DAY X 31 DAYS =		LBS
YEARLY TOTAL MILK PRODUCED =				LBS

TOTAL VALUE OF MILK PRODUCED FOR THE YEAR

| | LBS X $ | | VALUE PER LBS = | |

GOAT INFORMATION

NAME		BUCK	DOE
BREED		BIRTH DATE:	
DATE ACQUIRED:	HOW ACQUIRED: BORN ON FARM ☐	PURCHASED ☐	LEASED ☐
COLORS / IDENTIFYING MARKS:			
PURPOSE: MILK ☐	MEAT ☐ PET ☐	OTHER ☐	
NOTES			

PEDIGREE CHART

- SIRE
  - GRAND SIRE
    - GREAT GRAND SIRE
    - GREAT GRAND DAM
  - GRAND DAM
    - GREAT GRAND SIRE
    - GREAT GRAND DAM
- DAM
  - GRAND SIRE
    - GREAT GRAND SIRE
    - GREAT GRAND DAM
  - GRAND DAM
    - GREAT GRAND SIRE
    - GREAT GRAND DAM

MEDICAL INFORMATION

INJURY OR ILLNESS

DATE	DESCRIPTION OR NATURE OF ILLNESS	TREATMENT

PARASITE CONTROL

DATE	METHOD OR DEWORMER	DATE	METHOD OR DEWORMER

TESTING RECORD

DATE	TEST PERFORMED (CAE, CL, TB...)	RESULT	DATE	TEST PERFORMED (CAE, CL, TB...)	RESULT

VACCINATION & SUPPLEMENT RECORD

DATE	TARGET DISEASE	DRUG OR SUPPLEMENT USED	DOSAGE	RESULTS

DOE'S KIDDING RECORD

DOE'S NAME:

DATE BREED	KIDDING DATE	# OF KIDS	SEX D/B	NAME OF KID	SIRE OF KID	WEIGHT	TATTOO

BUCK'S RECORD OF PROGENY

BUCK'S NAME:	

YEAR	BRED TO	KIDS	DOE/BUCK

GOAT RECORD

GOAT'S NAME:		IDENTIFICATION:	
BREED:	DATE OF BIRTH:		DATE OF WEANED:

WEIGHT (POUNDS)

BIRTH	JAN	FEB	MAR	APR	MAY	JUN	JUL	AUG	SEP	OCT	NOV	DEC	FINAL

FEED RECORD

	JAN	FEB	MAR	APR	MAY	JUN	JUL	AUG	SEP	OCT	NOV	DEC	TOTAL
GRAIN													
FIELD													
PASTURE													

MILK PRODUCTION

DOE'S NAME:		IDENTIFICATION:	
BREED:	DATE OF BIRTH:	KIDDING DATE:	

JANUARY		AVERAGE LBS / DAY X 31 DAYS =		LBS
FEBRUARY		AVERAGE LBS / DAY X 31 DAYS =		LBS
MARCH		AVERAGE LBS / DAY X 31 DAYS =		LBS
APRIL		AVERAGE LBS / DAY X 31 DAYS =		LBS
MAY		AVERAGE LBS / DAY X 31 DAYS =		LBS
JUNE		AVERAGE LBS / DAY X 31 DAYS =		LBS
JULY		AVERAGE LBS / DAY X 31 DAYS =		LBS
AUGUST		AVERAGE LBS / DAY X 31 DAYS =		LBS
SEPTEMBER		AVERAGE LBS / DAY X 31 DAYS =		LBS
OCTOBER		AVERAGE LBS / DAY X 31 DAYS =		LBS
NOVEMBER		AVERAGE LBS / DAY X 31 DAYS =		LBS
DECEMBER		AVERAGE LBS / DAY X 31 DAYS =		LBS
YEARLY TOTAL MILK PRODUCED =				LBS

TOTAL VALUE OF MILK PRODUCED FOR THE YEAR

| | LBS X $ | | VALUE PER LBS = | |

GOAT INFORMATION

NAME		BUCK	DOE
BREED		BIRTH DATE:	
DATE ACQUIRED:	HOW ACQUIRED: BORN ON FARM ☐ PURCHASED ☐ LEASED ☐		
COLORS / IDENTIFYING MARKS:			
PURPOSE:	MILK ☐ MEAT ☐ PET ☐ OTHER ☐		
NOTES			

PEDIGREE CHART

- SIRE
  - GRAND SIRE
    - GREAT GRAND SIRE
    - GREAT GRAND DAM
  - GRAND DAM
    - GREAT GRAND SIRE
    - GREAT GRAND DAM
- DAM
  - GRAND SIRE
    - GREAT GRAND SIRE
    - GREAT GRAND DAM
  - GRAND DAM
    - GREAT GRAND SIRE
    - GREAT GRAND DAM

MEDICAL INFORMATION

INJURY OR ILLNESS

DATE	DESCRIPTION OR NATURE OF ILLNESS	TREATMENT

PARASITE CONTROL

DATE	METHOD OR DEWORMER	DATE	METHOD OR DEWORMER

TESTING RECORD

DATE	TEST PERFORMED (CAE, CL, TB...)	RESULT	DATE	TEST PERFORMED (CAE, CL, TB...)	RESULT

VACCINATION & SUPPLEMENT RECORD

DATE	TARGET DISEASE	DRUG OR SUPPLEMENT USED	DOSAGE	RESULTS

DOE'S KIDDING RECORD

DOE'S NAME:

DATE BREED	KIDDING DATE	# OF KIDS	SEX D/B	NAME OF KID	SIRE OF KID	WEIGHT	TATTOO

BUCK'S RECORD OF PROGENY

BUCK'S NAME:			

YEAR	BRED TO	KIDS	DOE/BUCK

GOAT RECORD

GOAT'S NAME:	IDENTIFICATION:
BREED:	DATE OF BIRTH:	DATE OF WEANED:

WEIGHT (POUNDS)

BIRTH	JAN	FEB	MAR	APR	MAY	JUN	JUL	AUG	SEP	OCT	NOV	DEC	FINAL

FEED RECORD

	JAN	FEB	MAR	APR	MAY	JUN	JUL	AUG	SEP	OCT	NOV	DEC	TOTAL
GRAIN													
FIELD													
PASTURE													

MILK PRODUCTION

DOE'S NAME:		IDENTIFICATION:	
BREED:	DATE OF BIRTH:	KIDDING DATE:	

JANUARY		AVERAGE LBS / DAY X 31 DAYS =		LBS
FEBRUARY		AVERAGE LBS / DAY X 31 DAYS =		LBS
MARCH		AVERAGE LBS / DAY X 31 DAYS =		LBS
APRIL		AVERAGE LBS / DAY X 31 DAYS =		LBS
MAY		AVERAGE LBS / DAY X 31 DAYS =		LBS
JUNE		AVERAGE LBS / DAY X 31 DAYS =		LBS
JULY		AVERAGE LBS / DAY X 31 DAYS =		LBS
AUGUST		AVERAGE LBS / DAY X 31 DAYS =		LBS
SEPTEMBER		AVERAGE LBS / DAY X 31 DAYS =		LBS
OCTOBER		AVERAGE LBS / DAY X 31 DAYS =		LBS
NOVEMBER		AVERAGE LBS / DAY X 31 DAYS =		LBS
DECEMBER		AVERAGE LBS / DAY X 31 DAYS =		LBS
YEARLY TOTAL MILK PRODUCED =				LBS

TOTAL VALUE OF MILK PRODUCED FOR THE YEAR

| | LBS X $ | | VALUE PER LBS = | |

GOAT INFORMATION

NAME		BUCK	DOE
BREED		BIRTH DATE:	
DATE ACQUIRED:	HOW ACQUIRED: BORN ON FARM ☐	PURCHASED ☐	LEASED ☐
COLORS / IDENTIFYING MARKS:			
PURPOSE: MILK ☐	MEAT ☐ PET ☐	OTHER ☐	
NOTES			

PEDIGREE CHART

- SIRE
  - GRAND SIRE
    - GREAT GRAND SIRE
    - GREAT GRAND DAM
  - GRAND DAM
    - GREAT GRAND SIRE
    - GREAT GRAND DAM
- DAM
  - GRAND SIRE
    - GREAT GRAND SIRE
    - GREAT GRAND DAM
  - GRAND DAM
    - GREAT GRAND SIRE
    - GREAT GRAND DAM

MEDICAL INFORMATION

INJURY OR ILLNESS

DATE	DESCRIPTION OR NATURE OF ILLNESS	TREATMENT

PARASITE CONTROL

DATE	METHOD OR DEWORMER	DATE	METHOD OR DEWORMER

TESTING RECORD

DATE	TEST PERFORMED (CAE, CL, TB...)	RESULT	DATE	TEST PERFORMED (CAE, CL, TB...)	RESULT

VACCINATION & SUPPLEMENT RECORD

DATE	TARGET DISEASE	DRUG OR SUPPLEMENT USED	DOSAGE	RESULTS

DOE'S KIDDING RECORD

DOE'S NAME:

DATE BREED	KIDDING DATE	# OF KIDS	SEX D/B	NAME OF KID	SIRE OF KID	WEIGHT	TATTOO

BUCK'S RECORD OF PROGENY

BUCK'S NAME:	

YEAR	BRED TO	KIDS	DOE/BUCK

GOAT RECORD

GOAT'S NAME:		IDENTIFICATION:
BREED:	DATE OF BIRTH:	DATE OF WEANED:

WEIGHT (POUNDS)

BIRTH	JAN	FEB	MAR	APR	MAY	JUN	JUL	AUG	SEP	OCT	NOV	DEC	FINAL

FEED RECORD

	JAN	FEB	MAR	APR	MAY	JUN	JUL	AUG	SEP	OCT	NOV	DEC	TOTAL
GRAIN													
FIELD													
PASTURE													

MILK PRODUCTION

DOE'S NAME:		IDENTIFICATION:	
BREED:	DATE OF BIRTH:	KIDDING DATE:	

JANUARY		AVERAGE LBS / DAY X 31 DAYS =		LBS
FEBRUARY		AVERAGE LBS / DAY X 31 DAYS =		LBS
MARCH		AVERAGE LBS / DAY X 31 DAYS =		LBS
APRIL		AVERAGE LBS / DAY X 31 DAYS =		LBS
MAY		AVERAGE LBS / DAY X 31 DAYS =		LBS
JUNE		AVERAGE LBS / DAY X 31 DAYS =		LBS
JULY		AVERAGE LBS / DAY X 31 DAYS =		LBS
AUGUST		AVERAGE LBS / DAY X 31 DAYS =		LBS
SEPTEMBER		AVERAGE LBS / DAY X 31 DAYS =		LBS
OCTOBER		AVERAGE LBS / DAY X 31 DAYS =		LBS
NOVEMBER		AVERAGE LBS / DAY X 31 DAYS =		LBS
DECEMBER		AVERAGE LBS / DAY X 31 DAYS =		LBS
YEARLY TOTAL MILK PRODUCED =				LBS

TOTAL VALUE OF MILK PRODUCED FOR THE YEAR

| | LBS X $ | | VALUE PER LBS = | |

GOAT INFORMATION

PHOTO

NAME		BUCK	DOE
BREED		BIRTH DATE:	
DATE ACQUIRED:	HOW ACQUIRED: BORN ON FARM ☐ PURCHASED ☐ LEASED ☐		
COLORS / IDENTIFYING MARKS:			
PURPOSE:	MILK ☐ MEAT ☐ PET ☐ OTHER ☐		
NOTES			

PEDIGREE CHART

SIRE	GRAND SIRE	GREAT GRAND SIRE	
		GREAT GRAND DAM	
	GRAND DAM	GREAT GRAND SIRE	
		GREAT GRAND DAM	
DAM	GRAND SIRE	GREAT GRAND SIRE	
		GREAT GRAND DAM	
	GRAND DAM	GREAT GRAND SIRE	
		GREAT GRAND DAM	

MEDICAL INFORMATION

INJURY OR ILLNESS

DATE	DESCRIPTION OR NATURE OF ILLNESS	TREATMENT

PARASITE CONTROL

DATE	METHOD OR DEWORMER	DATE	METHOD OR DEWORMER

TESTING RECORD

DATE	TEST PERFORMED (CAE, CL, TB...)	RESULT	DATE	TEST PERFORMED (CAE, CL, TB...)	RESULT

VACCINATION & SUPPLEMENT RECORD

DATE	TARGET DISEASE	DRUG OR SUPPLEMENT USED	DOSAGE	RESULTS

DOE'S KIDDING RECORD

DOE'S NAME:	

DATE BREED	KIDDING DATE	# OF KIDS	SEX D/B	NAME OF KID	SIRE OF KID	WEIGHT	TATTOO

BUCK'S RECORD OF PROGENY

BUCK'S NAME:	

YEAR	BRED TO	KIDS	DOE/BUCK

GOAT RECORD

GOAT'S NAME:	IDENTIFICATION:
BREED:	DATE OF BIRTH:	DATE OF WEANED:

WEIGHT (POUNDS)

BIRTH	JAN	FEB	MAR	APR	MAY	JUN	JUL	AUG	SEP	OCT	NOV	DEC	FINAL

FEED RECORD

	JAN	FEB	MAR	APR	MAY	JUN	JUL	AUG	SEP	OCT	NOV	DEC	TOTAL
GRAIN													
FIELD													
PASTURE													

MILK PRODUCTION

DOE'S NAME:		IDENTIFICATION:	
BREED:	DATE OF BIRTH:	KIDDING DATE:	

JANUARY		AVERAGE LBS / DAY X 31 DAYS =		LBS
FEBRUARY		AVERAGE LBS / DAY X 31 DAYS =		LBS
MARCH		AVERAGE LBS / DAY X 31 DAYS =		LBS
APRIL		AVERAGE LBS / DAY X 31 DAYS =		LBS
MAY		AVERAGE LBS / DAY X 31 DAYS =		LBS
JUNE		AVERAGE LBS / DAY X 31 DAYS =		LBS
JULY		AVERAGE LBS / DAY X 31 DAYS =		LBS
AUGUST		AVERAGE LBS / DAY X 31 DAYS =		LBS
SEPTEMBER		AVERAGE LBS / DAY X 31 DAYS =		LBS
OCTOBER		AVERAGE LBS / DAY X 31 DAYS =		LBS
NOVEMBER		AVERAGE LBS / DAY X 31 DAYS =		LBS
DECEMBER		AVERAGE LBS / DAY X 31 DAYS =		LBS
YEARLY TOTAL MILK PRODUCED =				LBS

TOTAL VALUE OF MILK PRODUCED FOR THE YEAR

| | LBS X $ | | VALUE PER LBS = | |

GOAT INFORMATION

NAME		BUCK	DOE
BREED		BIRTH DATE:	
DATE ACQUIRED:	HOW ACQUIRED: BORN ON FARM ☐ PURCHASED ☐ LEASED ☐		
COLORS / IDENTIFYING MARKS:			
PURPOSE: MILK ☐ MEAT ☐	PET ☐	OTHER ☐	
NOTES			

PEDIGREE CHART

- SIRE
  - GRAND SIRE
    - GREAT GRAND SIRE
    - GREAT GRAND DAM
  - GRAND DAM
    - GREAT GRAND SIRE
    - GREAT GRAND DAM
- DAM
  - GRAND SIRE
    - GREAT GRAND SIRE
    - GREAT GRAND DAM
  - GRAND DAM
    - GREAT GRAND SIRE
    - GREAT GRAND DAM

MEDICAL INFORMATION

INJURY OR ILLNESS

DATE	DESCRIPTION OR NATURE OF ILLNESS	TREATMENT

PARASITE CONTROL

DATE	METHOD OR DEWORMER	DATE	METHOD OR DEWORMER

TESTING RECORD

DATE	TEST PERFORMED (CAE, CL, TB...)	RESULT	DATE	TEST PERFORMED (CAE, CL, TB...)	RESULT

VACCINATION & SUPPLEMENT RECORD

DATE	TARGET DISEASE	DRUG OR SUPPLEMENT USED	DOSAGE	RESULTS

DOE'S KIDDING RECORD

DOE'S NAME:

DATE BREED	KIDDING DATE	# OF KIDS	SEX D/B	NAME OF KID	SIRE OF KID	WEIGHT	TATTOO

BUCK'S RECORD OF PROGENY

BUCK'S NAME:	

YEAR	BRED TO	KIDS	DOE/BUCK

GOAT RECORD

GOAT'S NAME:	IDENTIFICATION:
BREED:	DATE OF BIRTH:	DATE OF WEANED:

WEIGHT (POUNDS)

BIRTH	JAN	FEB	MAR	APR	MAY	JUN	JUL	AUG	SEP	OCT	NOV	DEC	FINAL

FEED RECORD

	JAN	FEB	MAR	APR	MAY	JUN	JUL	AUG	SEP	OCT	NOV	DEC	TOTAL
GRAIN													
FIELD													
PASTURE													

MILK PRODUCTION

DOE'S NAME:		IDENTIFICATION:	
BREED:	DATE OF BIRTH:	KIDDING DATE:	

JANUARY		AVERAGE LBS / DAY X 31 DAYS =		LBS
FEBRUARY		AVERAGE LBS / DAY X 31 DAYS =		LBS
MARCH		AVERAGE LBS / DAY X 31 DAYS =		LBS
APRIL		AVERAGE LBS / DAY X 31 DAYS =		LBS
MAY		AVERAGE LBS / DAY X 31 DAYS =		LBS
JUNE		AVERAGE LBS / DAY X 31 DAYS =		LBS
JULY		AVERAGE LBS / DAY X 31 DAYS =		LBS
AUGUST		AVERAGE LBS / DAY X 31 DAYS =		LBS
SEPTEMBER		AVERAGE LBS / DAY X 31 DAYS =		LBS
OCTOBER		AVERAGE LBS / DAY X 31 DAYS =		LBS
NOVEMBER		AVERAGE LBS / DAY X 31 DAYS =		LBS
DECEMBER		AVERAGE LBS / DAY X 31 DAYS =		LBS
YEARLY TOTAL MILK PRODUCED =				LBS
TOTAL VALUE OF MILK PRODUCED FOR THE YEAR				
	LBS X $		VALUE PER LBS =	

GOAT INFORMATION

PHOTO

NAME		BUCK	DOE
BREED		BIRTH DATE:	
DATE ACQUIRED:	HOW ACQUIRED: BORN ON FARM ☐ PURCHASED ☐ LEASED ☐		
COLORS / IDENTIFYING MARKS:			
PURPOSE:	MILK ☐ MEAT ☐ PET ☐	OTHER ☐	
NOTES			

PEDIGREE CHART

- SIRE
  - GRAND SIRE
    - GREAT GRAND SIRE
    - GREAT GRAND DAM
  - GRAND DAM
    - GREAT GRAND SIRE
    - GREAT GRAND DAM
- DAM
  - GRAND SIRE
    - GREAT GRAND SIRE
    - GREAT GRAND DAM
  - GRAND DAM
    - GREAT GRAND SIRE
    - GREAT GRAND DAM

MEDICAL INFORMATION

INJURY OR ILLNESS

DATE	DESCRIPTION OR NATURE OF ILLNESS	TREATMENT

PARASITE CONTROL

DATE	METHOD OR DEWORMER	DATE	METHOD OR DEWORMER

TESTING RECORD

DATE	TEST PERFORMED (CAE, CL, TB...)	RESULT	DATE	TEST PERFORMED (CAE, CL, TB...)	RESULT

VACCINATION & SUPPLEMENT RECORD

DATE	TARGET DISEASE	DRUG OR SUPPLEMENT USED	DOSAGE	RESULTS

DOE'S KIDDING RECORD

DOE'S NAME:

DATE BREED	KIDDING DATE	# OF KIDS	SEX D/B	NAME OF KID	SIRE OF KID	WEIGHT	TATTOO

BUCK'S RECORD OF PROGENY

BUCK'S NAME:	

YEAR	BRED TO	KIDS	DOE/BUCK

GOAT RECORD

GOAT'S NAME:		IDENTIFICATION:	
BREED:	DATE OF BIRTH:		DATE OF WEANED:

WEIGHT (POUNDS)

BIRTH	JAN	FEB	MAR	APR	MAY	JUN	JUL	AUG	SEP	OCT	NOV	DEC	FINAL

FEED RECORD

	JAN	FEB	MAR	APR	MAY	JUN	JUL	AUG	SEP	OCT	NOV	DEC	TOTAL
GRAIN													
FIELD													
PASTURE													

MILK PRODUCTION

DOE'S NAME:		IDENTIFICATION:	
BREED:	DATE OF BIRTH:	KIDDING DATE:	

JANUARY		AVERAGE LBS / DAY X 31 DAYS =		LBS
FEBRUARY		AVERAGE LBS / DAY X 31 DAYS =		LBS
MARCH		AVERAGE LBS / DAY X 31 DAYS =		LBS
APRIL		AVERAGE LBS / DAY X 31 DAYS =		LBS
MAY		AVERAGE LBS / DAY X 31 DAYS =		LBS
JUNE		AVERAGE LBS / DAY X 31 DAYS =		LBS
JULY		AVERAGE LBS / DAY X 31 DAYS =		LBS
AUGUST		AVERAGE LBS / DAY X 31 DAYS =		LBS
SEPTEMBER		AVERAGE LBS / DAY X 31 DAYS =		LBS
OCTOBER		AVERAGE LBS / DAY X 31 DAYS =		LBS
NOVEMBER		AVERAGE LBS / DAY X 31 DAYS =		LBS
DECEMBER		AVERAGE LBS / DAY X 31 DAYS =		LBS
YEARLY TOTAL MILK PRODUCED =				LBS

TOTAL VALUE OF MILK PRODUCED FOR THE YEAR

| | LBS X $ | | VALUE PER LBS = | |

GOAT INFORMATION

NAME		BUCK	DOE
BREED		BIRTH DATE:	
DATE ACQUIRED:	HOW ACQUIRED: BORN ON FARM ☐ PURCHASED ☐ LEASED ☐		
COLORS / IDENTIFYING MARKS:			
PURPOSE:	MILK ☐ MEAT ☐ PET ☐	OTHER ☐	
NOTES			

PEDIGREE CHART

- SIRE
  - GRAND SIRE
    - GREAT GRAND SIRE
    - GREAT GRAND DAM
  - GRAND DAM
    - GREAT GRAND SIRE
    - GREAT GRAND DAM
- DAM
  - GRAND SIRE
    - GREAT GRAND SIRE
    - GREAT GRAND DAM
  - GRAND DAM
    - GREAT GRAND SIRE
    - GREAT GRAND DAM

MEDICAL INFORMATION

INJURY OR ILLNESS

DATE	DESCRIPTION OR NATURE OF ILLNESS	TREATMENT

PARASITE CONTROL

DATE	METHOD OR DEWORMER	DATE	METHOD OR DEWORMER

TESTING RECORD

DATE	TEST PERFORMED (CAE, CL, TB...)	RESULT	DATE	TEST PERFORMED (CAE, CL, TB...)	RESULT

VACCINATION & SUPPLEMENT RECORD

DATE	TARGET DISEASE	DRUG OR SUPPLEMENT USED	DOSAGE	RESULTS

DOE'S KIDDING RECORD

DOE'S NAME:

DATE BREED	KIDDING DATE	# OF KIDS	SEX D/B	NAME OF KID	SIRE OF KID	WEIGHT	TATTOO

BUCK'S RECORD OF PROGENY

BUCK'S NAME:	

YEAR	BRED TO	KIDS	DOE/BUCK

GOAT RECORD

GOAT'S NAME:	IDENTIFICATION:
BREED:	DATE OF BIRTH:	DATE OF WEANED:

WEIGHT (POUNDS)

BIRTH	JAN	FEB	MAR	APR	MAY	JUN	JUL	AUG	SEP	OCT	NOV	DEC	FINAL

FEED RECORD

	JAN	FEB	MAR	APR	MAY	JUN	JUL	AUG	SEP	OCT	NOV	DEC	TOTAL
GRAIN													
FIELD													
PASTURE													

MILK PRODUCTION

DOE'S NAME:		IDENTIFICATION:	
BREED:	DATE OF BIRTH:	KIDDING DATE:	

JANUARY		AVERAGE LBS / DAY X 31 DAYS =		LBS
FEBRUARY		AVERAGE LBS / DAY X 31 DAYS =		LBS
MARCH		AVERAGE LBS / DAY X 31 DAYS =		LBS
APRIL		AVERAGE LBS / DAY X 31 DAYS =		LBS
MAY		AVERAGE LBS / DAY X 31 DAYS =		LBS
JUNE		AVERAGE LBS / DAY X 31 DAYS =		LBS
JULY		AVERAGE LBS / DAY X 31 DAYS =		LBS
AUGUST		AVERAGE LBS / DAY X 31 DAYS =		LBS
SEPTEMBER		AVERAGE LBS / DAY X 31 DAYS =		LBS
OCTOBER		AVERAGE LBS / DAY X 31 DAYS =		LBS
NOVEMBER		AVERAGE LBS / DAY X 31 DAYS =		LBS
DECEMBER		AVERAGE LBS / DAY X 31 DAYS =		LBS
YEARLY TOTAL MILK PRODUCED =				LBS

TOTAL VALUE OF MILK PRODUCED FOR THE YEAR				
	LBS X $		VALUE PER LBS =	

GOAT INFORMATION

NAME	BUCK	DOE
BREED	BIRTH DATE:	
DATE ACQUIRED:	HOW ACQUIRED: BORN ON FARM ☐ PURCHASED ☐ LEASED ☐	
COLORS / IDENTIFYING MARKS:		
PURPOSE: MILK ☐ MEAT ☐ PET ☐ OTHER ☐		
NOTES		

PEDIGREE CHART

- SIRE
  - GRAND SIRE
    - GREAT GRAND SIRE
    - GREAT GRAND DAM
  - GRAND DAM
    - GREAT GRAND SIRE
    - GREAT GRAND DAM
- DAM
  - GRAND SIRE
    - GREAT GRAND SIRE
    - GREAT GRAND DAM
  - GRAND DAM
    - GREAT GRAND SIRE
    - GREAT GRAND DAM

MEDICAL INFORMATION

INJURY OR ILLNESS

DATE	DESCRIPTION OR NATURE OF ILLNESS	TREATMENT

PARASITE CONTROL

DATE	METHOD OR DEWORMER	DATE	METHOD OR DEWORMER

TESTING RECORD

DATE	TEST PERFORMED (CAE, CL, TB...)	RESULT	DATE	TEST PERFORMED (CAE, CL, TB...)	RESULT

VACCINATION & SUPPLEMENT RECORD

DATE	TARGET DISEASE	DRUG OR SUPPLEMENT USED	DOSAGE	RESULTS

DOE'S KIDDING RECORD

DOE'S NAME:

DATE BREED	KIDDING DATE	# OF KIDS	SEX D/B	NAME OF KID	SIRE OF KID	WEIGHT	TATTOO

BUCK'S RECORD OF PROGENY

BUCK'S NAME:	

YEAR	BRED TO	KIDS	DOE/BUCK

GOAT RECORD

GOAT'S NAME:		IDENTIFICATION:	
BREED:	DATE OF BIRTH:		DATE OF WEANED:

WEIGHT (POUNDS)

BIRTH	JAN	FEB	MAR	APR	MAY	JUN	JUL	AUG	SEP	OCT	NOV	DEC	FINAL

FEED RECORD

	JAN	FEB	MAR	APR	MAY	JUN	JUL	AUG	SEP	OCT	NOV	DEC	TOTAL
GRAIN													
FIELD													
PASTURE													

MILK PRODUCTION

DOE'S NAME:		IDENTIFICATION:	
BREED:	DATE OF BIRTH:	KIDDING DATE:	

JANUARY		AVERAGE LBS / DAY X 31 DAYS =		LBS
FEBRUARY		AVERAGE LBS / DAY X 31 DAYS =		LBS
MARCH		AVERAGE LBS / DAY X 31 DAYS =		LBS
APRIL		AVERAGE LBS / DAY X 31 DAYS =		LBS
MAY		AVERAGE LBS / DAY X 31 DAYS =		LBS
JUNE		AVERAGE LBS / DAY X 31 DAYS =		LBS
JULY		AVERAGE LBS / DAY X 31 DAYS =		LBS
AUGUST		AVERAGE LBS / DAY X 31 DAYS =		LBS
SEPTEMBER		AVERAGE LBS / DAY X 31 DAYS =		LBS
OCTOBER		AVERAGE LBS / DAY X 31 DAYS =		LBS
NOVEMBER		AVERAGE LBS / DAY X 31 DAYS =		LBS
DECEMBER		AVERAGE LBS / DAY X 31 DAYS =		LBS
YEARLY TOTAL MILK PRODUCED =				LBS

TOTAL VALUE OF MILK PRODUCED FOR THE YEAR

| | LBS X $ | | VALUE PER LBS = | |

GOAT INFORMATION

NAME		BUCK	DOE
BREED		BIRTH DATE:	
DATE ACQUIRED:	HOW ACQUIRED: BORN ON FARM ☐	PURCHASED ☐	LEASED ☐
COLORS / IDENTIFYING MARKS:			
PURPOSE: MILK ☐	MEAT ☐ PET ☐	OTHER ☐	
NOTES			

PEDIGREE CHART

- SIRE
  - GRAND SIRE
    - GREAT GRAND SIRE
    - GREAT GRAND DAM
  - GRAND DAM
    - GREAT GRAND SIRE
    - GREAT GRAND DAM
- DAM
  - GRAND SIRE
    - GREAT GRAND SIRE
    - GREAT GRAND DAM
  - GRAND DAM
    - GREAT GRAND SIRE
    - GREAT GRAND DAM

MEDICAL INFORMATION

INJURY OR ILLNESS

DATE	DESCRIPTION OR NATURE OF ILLNESS	TREATMENT

PARASITE CONTROL

DATE	METHOD OR DEWORMER	DATE	METHOD OR DEWORMER

TESTING RECORD

DATE	TEST PERFORMED (CAE, CL, TB...)	RESULT	DATE	TEST PERFORMED (CAE, CL, TB...)	RESULT

VACCINATION & SUPPLEMENT RECORD

DATE	TARGET DISEASE	DRUG OR SUPPLEMENT USED	DOSAGE	RESULTS

DOE'S KIDDING RECORD

DOE'S NAME:

DATE BREED	KIDDING DATE	# OF KIDS	SEX D/B	NAME OF KID	SIRE OF KID	WEIGHT	TATTOO

BUCK'S RECORD OF PROGENY

BUCK'S NAME:	

YEAR	BRED TO	KIDS	DOE/BUCK

GOAT RECORD

GOAT'S NAME:		IDENTIFICATION:
BREED:	DATE OF BIRTH:	DATE OF WEANED:

WEIGHT (POUNDS)

BIRTH	JAN	FEB	MAR	APR	MAY	JUN	JUL	AUG	SEP	OCT	NOV	DEC	FINAL

FEED RECORD

	JAN	FEB	MAR	APR	MAY	JUN	JUL	AUG	SEP	OCT	NOV	DEC	TOTAL
GRAIN													
FIELD													
PASTURE													

MILK PRODUCTION

DOE'S NAME:		IDENTIFICATION:	
BREED:	DATE OF BIRTH:	KIDDING DATE:	

JANUARY		AVERAGE LBS / DAY X 31 DAYS =		LBS
FEBRUARY		AVERAGE LBS / DAY X 31 DAYS =		LBS
MARCH		AVERAGE LBS / DAY X 31 DAYS =		LBS
APRIL		AVERAGE LBS / DAY X 31 DAYS =		LBS
MAY		AVERAGE LBS / DAY X 31 DAYS =		LBS
JUNE		AVERAGE LBS / DAY X 31 DAYS =		LBS
JULY		AVERAGE LBS / DAY X 31 DAYS =		LBS
AUGUST		AVERAGE LBS / DAY X 31 DAYS =		LBS
SEPTEMBER		AVERAGE LBS / DAY X 31 DAYS =		LBS
OCTOBER		AVERAGE LBS / DAY X 31 DAYS =		LBS
NOVEMBER		AVERAGE LBS / DAY X 31 DAYS =		LBS
DECEMBER		AVERAGE LBS / DAY X 31 DAYS =		LBS
YEARLY TOTAL MILK PRODUCED =				LBS

TOTAL VALUE OF MILK PRODUCED FOR THE YEAR

| | LBS X $ | | VALUE PER LBS = | |

GOAT INFORMATION

NAME		BUCK	DOE
BREED		BIRTH DATE:	
DATE ACQUIRED:	HOW ACQUIRED: BORN ON FARM ☐ PURCHASED ☐ LEASED ☐		
COLORS / IDENTIFYING MARKS:			
PURPOSE:	MILK ☐ MEAT ☐ PET ☐	OTHER ☐	
NOTES			

PEDIGREE CHART

- SIRE
  - GRAND SIRE
    - GREAT GRAND SIRE
    - GREAT GRAND DAM
  - GRAND DAM
    - GREAT GRAND SIRE
    - GREAT GRAND DAM
- DAM
  - GRAND SIRE
    - GREAT GRAND SIRE
    - GREAT GRAND DAM
  - GRAND DAM
    - GREAT GRAND SIRE
    - GREAT GRAND DAM

MEDICAL INFORMATION

INJURY OR ILLNESS

DATE	DESCRIPTION OR NATURE OF ILLNESS	TREATMENT

PARASITE CONTROL

DATE	METHOD OR DEWORMER	DATE	METHOD OR DEWORMER

TESTING RECORD

DATE	TEST PERFORMED (CAE, CL, TB...)	RESULT	DATE	TEST PERFORMED (CAE, CL, TB...)	RESULT

VACCINATION & SUPPLEMENT RECORD

DATE	TARGET DISEASE	DRUG OR SUPPLEMENT USED	DOSAGE	RESULTS

DOE'S KIDDING RECORD

DOE'S NAME:

DATE BREED	KIDDING DATE	# OF KIDS	SEX D/B	NAME OF KID	SIRE OF KID	WEIGHT	TATTOO

BUCK'S RECORD OF PROGENY

BUCK'S NAME:	

YEAR	BRED TO	KIDS	DOE/BUCK

GOAT RECORD

GOAT'S NAME:		IDENTIFICATION:
BREED:	DATE OF BIRTH:	DATE OF WEANED:

WEIGHT (POUNDS)

BIRTH	JAN	FEB	MAR	APR	MAY	JUN	JUL	AUG	SEP	OCT	NOV	DEC	FINAL

FEED RECORD

	JAN	FEB	MAR	APR	MAY	JUN	JUL	AUG	SEP	OCT	NOV	DEC	TOTAL
GRAIN													
FIELD													
PASTURE													

MILK PRODUCTION

DOE'S NAME:		IDENTIFICATION:	
BREED:	DATE OF BIRTH:	KIDDING DATE:	

JANUARY		AVERAGE LBS / DAY X 31 DAYS =		LBS
FEBRUARY		AVERAGE LBS / DAY X 31 DAYS =		LBS
MARCH		AVERAGE LBS / DAY X 31 DAYS =		LBS
APRIL		AVERAGE LBS / DAY X 31 DAYS =		LBS
MAY		AVERAGE LBS / DAY X 31 DAYS =		LBS
JUNE		AVERAGE LBS / DAY X 31 DAYS =		LBS
JULY		AVERAGE LBS / DAY X 31 DAYS =		LBS
AUGUST		AVERAGE LBS / DAY X 31 DAYS =		LBS
SEPTEMBER		AVERAGE LBS / DAY X 31 DAYS =		LBS
OCTOBER		AVERAGE LBS / DAY X 31 DAYS =		LBS
NOVEMBER		AVERAGE LBS / DAY X 31 DAYS =		LBS
DECEMBER		AVERAGE LBS / DAY X 31 DAYS =		LBS
YEARLY TOTAL MILK PRODUCED =				LBS

TOTAL VALUE OF MILK PRODUCED FOR THE YEAR

| | LBS X $ | | VALUE PER LBS = | |

GOAT INFORMATION

PHOTO

NAME		BUCK	DOE
BREED		BIRTH DATE:	
DATE ACQUIRED:	HOW ACQUIRED: BORN ON FARM ☐ PURCHASED ☐ LEASED ☐		
COLORS / IDENTIFYING MARKS:			
PURPOSE:	MILK ☐ MEAT ☐ PET ☐	OTHER ☐	
NOTES			

PEDIGREE CHART

- SIRE
  - GRAND SIRE
    - GREAT GRAND SIRE
    - GREAT GRAND DAM
  - GRAND DAM
    - GREAT GRAND SIRE
    - GREAT GRAND DAM
- DAM
  - GRAND SIRE
    - GREAT GRAND SIRE
    - GREAT GRAND DAM
  - GRAND DAM
    - GREAT GRAND SIRE
    - GREAT GRAND DAM

MEDICAL INFORMATION

INJURY OR ILLNESS

DATE	DESCRIPTION OR NATURE OF ILLNESS	TREATMENT

PARASITE CONTROL

DATE	METHOD OR DEWORMER	DATE	METHOD OR DEWORMER

TESTING RECORD

DATE	TEST PERFORMED (CAE, CL, TB...)	RESULT	DATE	TEST PERFORMED (CAE, CL, TB...)	RESULT

VACCINATION & SUPPLEMENT RECORD

DATE	TARGET DISEASE	DRUG OR SUPPLEMENT USED	DOSAGE	RESULTS

DOE'S KIDDING RECORD

DOE'S NAME:

DATE BREED	KIDDING DATE	# OF KIDS	SEX D/B	NAME OF KID	SIRE OF KID	WEIGHT	TATTOO

BUCK'S RECORD OF PROGENY

BUCK'S NAME:	

YEAR	BRED TO	KIDS	DOE/BUCK

GOAT RECORD

GOAT'S NAME:	IDENTIFICATION:
BREED:	DATE OF BIRTH:	DATE OF WEANED:

WEIGHT (POUNDS)

BIRTH	JAN	FEB	MAR	APR	MAY	JUN	JUL	AUG	SEP	OCT	NOV	DEC	FINAL

FEED RECORD

	JAN	FEB	MAR	APR	MAY	JUN	JUL	AUG	SEP	OCT	NOV	DEC	TOTAL
GRAIN													
FIELD													
PASTURE													

MILK PRODUCTION

DOE'S NAME:		IDENTIFICATION:		
BREED:	DATE OF BIRTH:		KIDDING DATE:	

JANUARY		AVERAGE LBS / DAY X 31 DAYS =		LBS
FEBRUARY		AVERAGE LBS / DAY X 31 DAYS =		LBS
MARCH		AVERAGE LBS / DAY X 31 DAYS =		LBS
APRIL		AVERAGE LBS / DAY X 31 DAYS =		LBS
MAY		AVERAGE LBS / DAY X 31 DAYS =		LBS
JUNE		AVERAGE LBS / DAY X 31 DAYS =		LBS
JULY		AVERAGE LBS / DAY X 31 DAYS =		LBS
AUGUST		AVERAGE LBS / DAY X 31 DAYS =		LBS
SEPTEMBER		AVERAGE LBS / DAY X 31 DAYS =		LBS
OCTOBER		AVERAGE LBS / DAY X 31 DAYS =		LBS
NOVEMBER		AVERAGE LBS / DAY X 31 DAYS =		LBS
DECEMBER		AVERAGE LBS / DAY X 31 DAYS =		LBS
YEARLY TOTAL MILK PRODUCED =				LBS

TOTAL VALUE OF MILK PRODUCED FOR THE YEAR

| | LBS X $ | | VALUE PER LBS = | |

GOAT INFORMATION

NAME	BUCK	DOE
BREED	BIRTH DATE:	
DATE ACQUIRED:	HOW ACQUIRED: BORN ON FARM ☐ PURCHASED ☐ LEASED ☐	
COLORS / IDENTIFYING MARKS:		
PURPOSE: MILK ☐ MEAT ☐ PET ☐ OTHER ☐		
NOTES		

PEDIGREE CHART

- SIRE
  - GRAND SIRE
    - GREAT GRAND SIRE
    - GREAT GRAND DAM
  - GRAND DAM
    - GREAT GRAND SIRE
    - GREAT GRAND DAM
- DAM
  - GRAND SIRE
    - GREAT GRAND SIRE
    - GREAT GRAND DAM
  - GRAND DAM
    - GREAT GRAND SIRE
    - GREAT GRAND DAM

MEDICAL INFORMATION

INJURY OR ILLNESS

DATE	DESCRIPTION OR NATURE OF ILLNESS	TREATMENT

PARASITE CONTROL

DATE	METHOD OR DEWORMER	DATE	METHOD OR DEWORMER

TESTING RECORD

DATE	TEST PERFORMED (CAE, CL, TB...)	RESULT	DATE	TEST PERFORMED (CAE, CL, TB...)	RESULT

VACCINATION & SUPPLEMENT RECORD

DATE	TARGET DISEASE	DRUG OR SUPPLEMENT USED	DOSAGE	RESULTS

DOE'S KIDDING RECORD

DOE'S NAME:	

DATE BRED	KIDDING DATE	# OF KIDS	SEX D/B	NAME OF KID	SIRE OF KID	WEIGHT	TATTOO

BUCK'S RECORD OF PROGENY

BUCK'S NAME:	

YEAR	BRED TO	KIDS	DOE/BUCK

GOAT RECORD

GOAT'S NAME:	IDENTIFICATION:
BREED:	DATE OF BIRTH:	DATE OF WEANED:

WEIGHT (POUNDS)

BIRTH	JAN	FEB	MAR	APR	MAY	JUN	JUL	AUG	SEP	OCT	NOV	DEC	FINAL

FEED RECORD

	JAN	FEB	MAR	APR	MAY	JUN	JUL	AUG	SEP	OCT	NOV	DEC	TOTAL
GRAIN													
FIELD													
PASTURE													

MILK PRODUCTION

DOE'S NAME:		IDENTIFICATION:	
BREED:	DATE OF BIRTH:	KIDDING DATE:	

JANUARY		AVERAGE LBS / DAY X 31 DAYS =		LBS
FEBRUARY		AVERAGE LBS / DAY X 31 DAYS =		LBS
MARCH		AVERAGE LBS / DAY X 31 DAYS =		LBS
APRIL		AVERAGE LBS / DAY X 31 DAYS =		LBS
MAY		AVERAGE LBS / DAY X 31 DAYS =		LBS
JUNE		AVERAGE LBS / DAY X 31 DAYS =		LBS
JULY		AVERAGE LBS / DAY X 31 DAYS =		LBS
AUGUST		AVERAGE LBS / DAY X 31 DAYS =		LBS
SEPTEMBER		AVERAGE LBS / DAY X 31 DAYS =		LBS
OCTOBER		AVERAGE LBS / DAY X 31 DAYS =		LBS
NOVEMBER		AVERAGE LBS / DAY X 31 DAYS =		LBS
DECEMBER		AVERAGE LBS / DAY X 31 DAYS =		LBS
YEARLY TOTAL MILK PRODUCED =				LBS

TOTAL VALUE OF MILK PRODUCED FOR THE YEAR

| | LBS X $ | | VALUE PER LBS = | |

GOAT INFORMATION

NAME		BUCK	DOE
BREED		BIRTH DATE:	
DATE ACQUIRED:	HOW ACQUIRED: BORN ON FARM ☐ PURCHASED ☐ LEASED ☐		
COLORS / IDENTIFYING MARKS:			
PURPOSE:	MILK ☐ MEAT ☐ PET ☐	OTHER ☐	
NOTES			

PEDIGREE CHART

- SIRE
  - GRAND SIRE
    - GREAT GRAND SIRE
    - GREAT GRAND DAM
  - GRAND DAM
    - GREAT GRAND SIRE
    - GREAT GRAND DAM
- DAM
  - GRAND SIRE
    - GREAT GRAND SIRE
    - GREAT GRAND DAM
  - GRAND DAM
    - GREAT GRAND SIRE
    - GREAT GRAND DAM

MEDICAL INFORMATION

INJURY OR ILLNESS

DATE	DESCRIPTION OR NATURE OF ILLNESS	TREATMENT

PARASITE CONTROL

DATE	METHOD OR DEWORMER	DATE	METHOD OR DEWORMER

TESTING RECORD

DATE	TEST PERFORMED (CAE, CL, TB...)	RESULT	DATE	TEST PERFORMED (CAE, CL, TB...)	RESULT

VACCINATION & SUPPLEMENT RECORD

DATE	TARGET DISEASE	DRUG OR SUPPLEMENT USED	DOSAGE	RESULTS

DOE'S KIDDING RECORD

DOE'S NAME:

DATE BREED	KIDDING DATE	# OF KIDS	SEX D/B	NAME OF KID	SIRE OF KID	WEIGHT	TATTOO

BUCK'S RECORD OF PROGENY

BUCK'S NAME:	

YEAR	BRED TO	KIDS	DOE/BUCK

GOAT RECORD

GOAT'S NAME:		IDENTIFICATION:	
BREED:	DATE OF BIRTH:		DATE OF WEANED:

WEIGHT (POUNDS)

BIRTH	JAN	FEB	MAR	APR	MAY	JUN	JUL	AUG	SEP	OCT	NOV	DEC	FINAL

FEED RECORD

	JAN	FEB	MAR	APR	MAY	JUN	JUL	AUG	SEP	OCT	NOV	DEC	TOTAL
GRAIN													
FIELD													
PASTURE													

MILK PRODUCTION

DOE'S NAME:		IDENTIFICATION:	
BREED:	DATE OF BIRTH:	KIDDING DATE:	

JANUARY		AVERAGE LBS / DAY X 31 DAYS =		LBS
FEBRUARY		AVERAGE LBS / DAY X 31 DAYS =		LBS
MARCH		AVERAGE LBS / DAY X 31 DAYS =		LBS
APRIL		AVERAGE LBS / DAY X 31 DAYS =		LBS
MAY		AVERAGE LBS / DAY X 31 DAYS =		LBS
JUNE		AVERAGE LBS / DAY X 31 DAYS =		LBS
JULY		AVERAGE LBS / DAY X 31 DAYS =		LBS
AUGUST		AVERAGE LBS / DAY X 31 DAYS =		LBS
SEPTEMBER		AVERAGE LBS / DAY X 31 DAYS =		LBS
OCTOBER		AVERAGE LBS / DAY X 31 DAYS =		LBS
NOVEMBER		AVERAGE LBS / DAY X 31 DAYS =		LBS
DECEMBER		AVERAGE LBS / DAY X 31 DAYS =		LBS
YEARLY TOTAL MILK PRODUCED =				LBS
TOTAL VALUE OF MILK PRODUCED FOR THE YEAR				
	LBS X $		VALUE PER LBS =	

GOAT INFORMATION

PHOTO

NAME	BUCK	DOE
BREED	BIRTH DATE:	
DATE ACQUIRED:	HOW ACQUIRED: BORN ON FARM ☐ PURCHASED ☐ LEASED ☐	
COLORS / IDENTIFYING MARKS:		
PURPOSE: MILK ☐ MEAT ☐ PET ☐ OTHER ☐		
NOTES		

PEDIGREE CHART

- SIRE
  - GRAND SIRE
    - GREAT GRAND SIRE
    - GREAT GRAND DAM
  - GRAND DAM
    - GREAT GRAND SIRE
    - GREAT GRAND DAM
- DAM
  - GRAND SIRE
    - GREAT GRAND SIRE
    - GREAT GRAND DAM
  - GRAND DAM
    - GREAT GRAND SIRE
    - GREAT GRAND DAM

MEDICAL INFORMATION

INJURY OR ILLNESS

DATE	DESCRIPTION OR NATURE OF ILLNESS	TREATMENT

PARASITE CONTROL

DATE	METHOD OR DEWORMER	DATE	METHOD OR DEWORMER

TESTING RECORD

DATE	TEST PERFORMED (CAE, CL, TB...)	RESULT	DATE	TEST PERFORMED (CAE, CL, TB...)	RESULT

VACCINATION & SUPPLEMENT RECORD

DATE	TARGET DISEASE	DRUG OR SUPPLEMENT USED	DOSAGE	RESULTS

DOE'S KIDDING RECORD

DOE'S NAME:

DATE BREED	KIDDING DATE	# OF KIDS	SEX D/B	NAME OF KID	SIRE OF KID	WEIGHT	TATTOO

BUCK'S RECORD OF PROGENY

BUCK'S NAME:	

YEAR	BRED TO	KIDS	DOE/BUCK

GOAT RECORD

GOAT'S NAME:	IDENTIFICATION:
BREED:	DATE OF BIRTH:	DATE OF WEANED:

WEIGHT (POUNDS)

BIRTH	JAN	FEB	MAR	APR	MAY	JUN	JUL	AUG	SEP	OCT	NOV	DEC	FINAL

FEED RECORD

	JAN	FEB	MAR	APR	MAY	JUN	JUL	AUG	SEP	OCT	NOV	DEC	TOTAL
GRAIN													
FIELD													
PASTURE													

MILK PRODUCTION

DOE'S NAME:		IDENTIFICATION:	
BREED:	DATE OF BIRTH:	KIDDING DATE:	

JANUARY		AVERAGE LBS / DAY X 31 DAYS =		LBS
FEBRUARY		AVERAGE LBS / DAY X 31 DAYS =		LBS
MARCH		AVERAGE LBS / DAY X 31 DAYS =		LBS
APRIL		AVERAGE LBS / DAY X 31 DAYS =		LBS
MAY		AVERAGE LBS / DAY X 31 DAYS =		LBS
JUNE		AVERAGE LBS / DAY X 31 DAYS =		LBS
JULY		AVERAGE LBS / DAY X 31 DAYS =		LBS
AUGUST		AVERAGE LBS / DAY X 31 DAYS =		LBS
SEPTEMBER		AVERAGE LBS / DAY X 31 DAYS =		LBS
OCTOBER		AVERAGE LBS / DAY X 31 DAYS =		LBS
NOVEMBER		AVERAGE LBS / DAY X 31 DAYS =		LBS
DECEMBER		AVERAGE LBS / DAY X 31 DAYS =		LBS
YEARLY TOTAL MILK PRODUCED =				LBS

TOTAL VALUE OF MILK PRODUCED FOR THE YEAR

| | LBS X $ | | VALUE PER LBS = | |

www.ingramcontent.com/pod-product-compliance
Lightning Source LLC
Chambersburg PA
CBHW081154070526
44583CB00021B/2840